The ABC of

MiCROFiBRE

Use and Best

Practice

I0402269

Joshua Jogo

HEATHROW GATEWAY PUBLISHING UK

Heathrow House, London

United Kingdom

Tel: UK: +447448522120

Email: joshuajogo@hotmail.com

This Book was presented to

....

for

....

by

....

....

on

....

Introduction

Microfibre clothes and mops trap up to 99.9% of bacteria and dirt while using less water (250m2 floor space requires just 1.5l) and eliminating the chemical residue on which bacteria thrive in hard-to-reach areas. Our easy-to-use range of mops, both damp and dry, ensures every space and surface can be cleaned efficiently and effectively. A simple click system enables contact free mop head changes and remover- risk of cross contamination, which is particularly beneficial for hospitals. Pre-prepared damp mops leave surfaces dry and eliminate the damp areas where bacteria breed. The unique trolley system enables well-organised and hygienic movement of pre-prepared mops and cloths. Ergonomic tool handles with a patented adjustable system facilitate better working positions and promote operator health Colour-coded tools can be designated for use on a specific task or in a particular area.

CHAPTER ONE

Historical background of Micrfibre

Microfiber is a type of material that is primarily used as an upholstery fabric because of it strong and durable nature, but it can also be found in other household products such as drapes, linens, and cleaning tools like mops. The material is less commonly used for clothing, because it is flammable, but it is often used for athletic attire because of its ability to wick moisture away from the body. Made primarily of petroleum waste products, the material makes use of waste that would otherwise fill up landfills.

Development

During the middle of the 20th century, the manufacturing of synthetic fibers began to expand into new areas. One of the breakthroughs was to take the sludge that was left over after oil had been refined and turn it into a synthetic fiber that could be used in upholstery;

the process of refining this waste product yielded a substance known as polypropylene, which could in turn be processed into a thin olefin fiber. Olefin fibers were ideal to use in the production of car upholstery, home and office carpeting, and even some draperies. Olefin fibers caught on in a big way during the 1970s with one company in particular, Hercules, Inc., producing their own branded form of olefin fiber, which was dubbed Herculon®.

Continued experimentation allowed for the use of polypropylene to develop an extremely fine fiber, today referred to as microfiber. This extremely thin, but surprisingly resilient, fiber could be used for a number of textile applications that the broader weaves of olefin fibers were simply not suited for because of their density. Today it is possible to transform what would have been waste into a highly-valued material.

Moisture-wicking

Microfiber has the excellent wicking properties, meaning it will absorb moisture and oils rather than allow them to set on the surface of the material. This makes it an ideal material for such things as footballs and basketballs, as the sweat from the players' hands will not make the ball slippery and harder to hold. The moisture-wicking characteristic also makes the material a popular choice for furniture, particularly couches, as most food and liquids can be easily wiped off before the furniture is stained.

Household Supplies

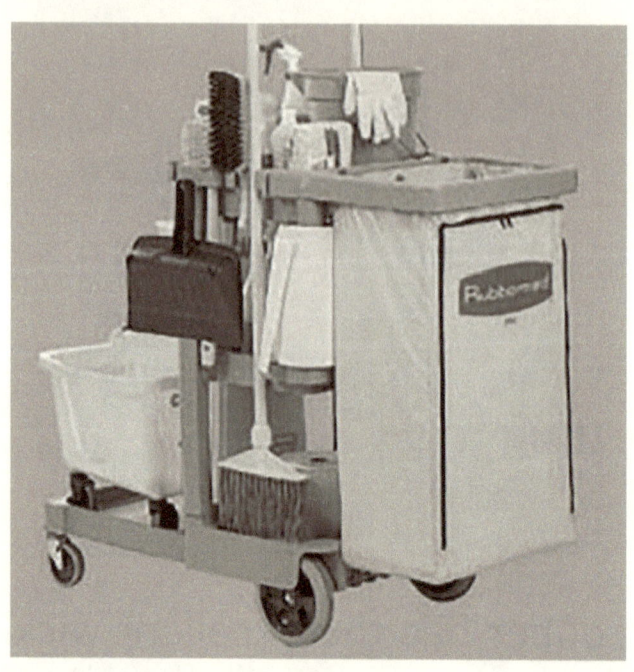

Microfibers are also used for various types of cloth where absorption of water is desirable; for instance, cleaning cloths used for dusting, cleaning glass, mopping, or detailing cars are often made of microfiber material. Most cloths made of this material do not leave behind residue of lint or dust, which makes them ideal for waxing a car; however, it is important to note that microfiber material will absorb dust and lint. A good idea is to wash the cloth after each use, to avoid any leftover residue that may be deposited the next time the cloth is used.

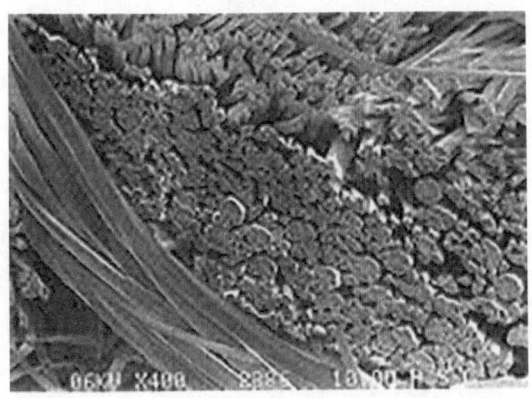

Linens

Microfiber is often used in other textile applications such as tablecloths, sheer draperies, and curtains. The stain-repelling ability of the microfiber blend in these types of products makes them very attractive to many homeowners, restaurant owners, and other types of business owners who prefer to use materials that are both good-looking and serviceable. Bath towels and hand towels are also popular when made in this material, as it is known to easily remove water and moisture.

Clothing

While the use of microfiber for clothing has been around for some years now, it is most often found in sports apparel due to its ability to wick away sweat. The material can also be found in medical compression socks, which are used to help patients increase blood flow. There are those that object to using the material for clothing because it is not a natural material and because it is flammable. Also, some people may feel that a garment made from microfiber material is not as comfortable as other fabric options.

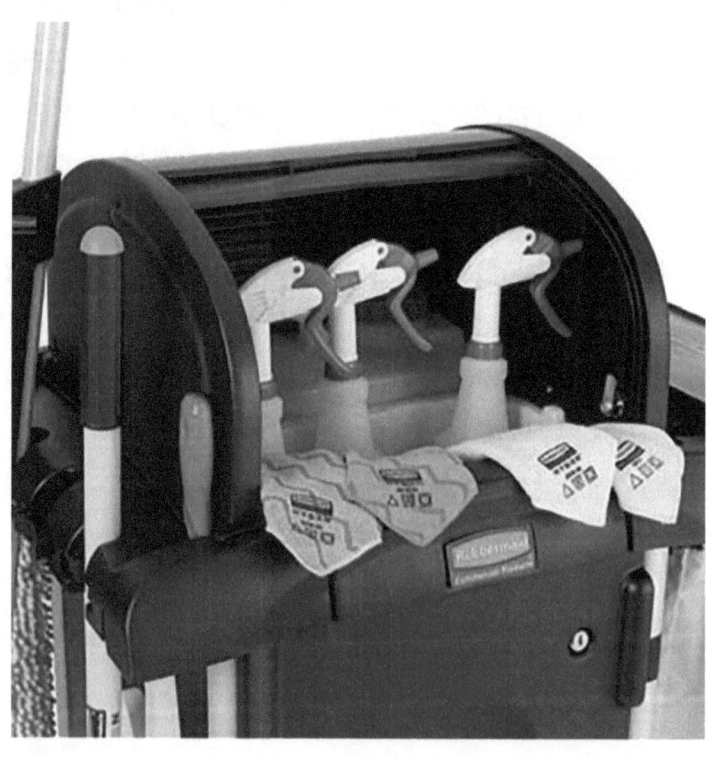

CHAPTER TWO

In General "Best Practice"

There are *3 key stages* that need to be considered when working with Micorofibre textiles

1. Preparation

2. Cleaning Process

3. Laundry and Logistics

Note: the process discussed in this chapter are based on the Vikan Microfibre cloth, mops and workstations and tested in shops cleaning at ASDA supermarkets in the UK.

Top Tip – *Preparation*

- Microfibre performs best damp not wet
- Always use the preparation jug and refer to Textile Guide
- Never Prep Mops on Workstations
- Check for standing water
- Check lids are secure, you will hear a click.
- Textiles should not drip or leave significant residue on floors or surfaces
-

Top Tip - *Cleaning Process*

- Use equipment correctly i.e. Right Equipment For The Right Job, correct prepping & storage
- Never use in warehouse
- Refer to How To Guide's
- Always clean from "Clean to Dirty"
- Change mops & cloths frequently – don't over soil the textiles

- Manage expectations in heavy soiled areas such as Bakery & Deli
-

Top Tip – *Laundry & Logistics*

- Correct laundry is essential
- Do not wash cloths & mops together
- Wash van cleaning textiles separately
- Ensure textiles are not over soiled vacuum or hand wash if necessary before placing in washer
- Shake cloths & mops over the rubbish sack before placing in the dirty boxes on the workstation
- Rotate stock and fold cloths with fold facing for easier counting and to speed up preparation

What would happen if and frequently asked Questions?

Prepping Textiles

Q. I am unsure how to prep my mops & cloths?

A. Refer to the Microfibre Textile Guide that should displayed in the cleaning room or speak to your Supervisor or SCM and ask them to explain

Q. I prepped both Blue & Red Microfibre cloths together e.g. 5 blue and 5 red

A. Don't worry once the all textiles have been laundered they are now hygienically clean so there is no risk of any cross contamination

Q. Can I prep Damp 43's with Damp 42's?
A. No, the water Measurement are different, the mops will either be too wet or too dry.

Q. What would happen if I did not use a measuring jug and just guessed the measurement?

A. This would not be an accurate measurement and the textiles will be too wet or too dry

Q. I have just started my shift and the Workstation has been pre-prepped incorrectly for me i.e. cloths and mops too wet and water lying in the bottom of the boxes, what should I do?

A. Squeeze excess water from textiles into sink area and then place to be tumble dried (separately) with the next available drying load's (they are still clean so no need to re-wash). Ask supervisor to re train colleague who had prepared

Q. I am in a hurry can I prep my mops in the Workstation boxes?

A. No, you must always prep mops on a flat surface so the water can be evenly divided over the mops, the Workstation boxes are slightly slanted so they will not be prepped correctly

Q. I pre-dampened a handful of Easyshine mop head and stored the same way as the Microfibre cloths on the Workstation?

A. You DO NOT have to pre-dampen the Easyshine mop heads and store in any boxes or Workstations, you only need to lightly dampen them as and when you need to use them, using a water spray bottle. Note: For storage you can keep the dry Easyshine mop heads in a Workstation boxes or Utility belt

Q. Can I use the Microfibre cloths dry and just damp spray when needed?

A. *No the cloths need to before-prepped in order to proactive*

Workstations

It is important that the workstations are set up exactly to the HTG

Q. Parts of my Workstation are coming loose what should I do?

A. *Check that all the tightening nuts are present and that the tool flex brackets are secure and tight before using as there may be a potential fall hazard, ask the handyman or SCM to tighten using a suitable "Cross Head Screwdriver"*

Q. Do I need to attach the squeegee blade to the lobby pan handle?

A. Yes, the lobby pan and brush comes fitted with two tools, a squeegee blade and a brush head each are attachable, one of these should be attached to the brush handle the other to the lobby pan handler. If you do not keep as a full set there is a strong possibility that one will go missing or get. Note: There are also coloured poppers (see below) that can be attached to the handle knob in the event that the Dustpan set is used frequently in colour coded areas

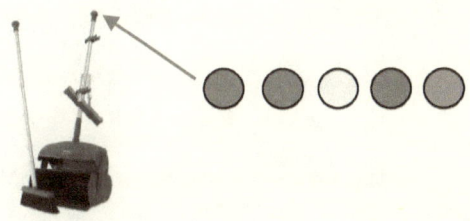

Q. On my Janitor Workstation where do I keep the Dry 43's?
A. There is a small plastic hook to the side of the Workstation where they hang or they can be stored in a mop box on the workstation

Q. If my Easyshine has no Velcro dots on the back; will I still be able to use it?

A. No, the Velcro dots have been designed to hold the Easyshine mop head in place to get a better clean. If required e mail Microfibresupport@vikan.co.uk for replacements dots.
Place the dots on the outer surface of the mop head (see below), attach the Easyshine mop head to the flat surface of the tool, fold over the cloth edges and attach to the Velcro dots on the outer surface or tool to secure.

Q. On my Workstation I have nowhere to keep / store breakages until I find time to dispose of them?

A. For storage space and damages you can use one of the spare empty boxes until you can take to the damage area. To keep the box clean place damages in a small blue spillage bag. You can also fit this small box into one of the larger boxes for storage

Q. Once my cloths are prepped and placed in the Workstation boxes is there anything else I need to do to keep them moist?

A. Nothing, if the cloths have been prepped correctly and the cloth Workstation box lids are securely closed they will stay damp throughout your shift

Q. How do I keep my Workstation multi bag bin holder clean? –
A. Inset a clear plastic bag to protect the inner liner

Q. I am a Janitor what size mop frames will I need on my Workstation?

A. You will need a 60cm mop frame and handle with dry 24 mop attached, a 40cm mop frame and handle to attach either a damp 42's or dry43's

Q. The cloth lids do not seem to close correctly there is a gap on each side, what do I do?

A. You may need to adjust the lid hinge. Remove the box from the Workstation, check that the plastic lid hinges is pushed up to the top of the hinge and the lid is secure and sitting correctly. Make sure box is on the workstation correctly (on one side of the box frame there is a slight indent groove for the boxes to sit correctly If this fails inform your Manager or Supervisor.)

Q. Where does the Working in Progress Sign and Lobby pan and brush go on my Workstation?

A. There are two hooks fitted onto the Workstation, one hanging from the bag holder for the Wet floor sign and the other is fitted to the other side box frame to hang your lobby pan and brush. Refer to WAGJLL / HTG Note: The dustpan hook can also hangs on the side of the bag holder.

Q. What would happen if I do not release the locking system on any of the mop heads and an Easyshine?

A. You must ALWAYS ensure that the locking system is off when you have finished your cleaning task, as the next person to use may not know it in a locked position and therefore there is a strong possibility of the equipment being broken or damaged.
Failure to do this will result in the locking mechanism braking.

Q. I have identified the Workstation that I use by wrapping tape around the handle, this way I know who used it last and what condition it was in when I got it back?

A. In theory the Workstations are not listed for any particular area of work, they are multi functional to be used in any department or area (except warehouse, car park and Asda Colleague CAYG trolleys) they are not colour coded or marked in anyway, so colleague should not wrap tape, string or anything else on the Workstations.

But we do understand the need to identify who used the Workstation last and had it been cleaned and stored correctly – so in this instance if you want to use the same Workstation you can discreetly use one of the colour coded stickers to place on the back frame behind the boxes/Waste Bag to identify (see below)

CHAPTER THREE

Dealing with Spillages & Breakages

Q. When dealing with a sales floor spillage the liquid/product is too much to go into the dustpan & squeegee, its seeps out the side

A. *Place a little Spill Aid into the lobby pan to help soak up liquid or for larger spillages machine clean. Tip: place a small blue spill sack into the lobby pan to keep clean*

Q. How do I deal with a spillage using the Microfibre textiles and equipment?

A. **The Dry 43 mop is insufficient when dealing with large spillages – the damp 43s are intended to deal with minor/small spillages. Large spillages should be dealt with as previously mentioned using the machine or spill aid**

Q. The dustpan does not hold liquid when dealing with spillages – when responding to liquid spillages such as wine, milk, pop etc

A. It is important that spill aid is used to soak up the liquid making it easier to deal. It may also be helpful if spill aid is sprinkled into the bottom of the dustpan. Note: Remember to clean the dustpan after dealing with spillages

Q. Do not use textiles to deal directly with spillages such as Shower Gels, Shampoo or washing detergents

A. No as this will transfer into the washing machine and cause excessive suds to form resulting in washing machine errors and failed wash cycles.

Deal with these spillages using the squeegee and dustpan, mops used to clean up any residue must be rinsed with clear water in the cleaning room before been transferred to the laundry bin

Q. All the Workstations are being used and I do not have one to clean the sales floor edges, what do I use?

A. Each Workstation has spare empty mop & cloth boxes, you can remove these boxes and use in your place of work i.e. link corridors, top of aisles if mopping sales floor edges or colleague area's i.e. offices, corridors and stairwell etc (Note: place boxes in a safe manner to minimise trips and falls)

Q. The floors are still looking dirty after initial clean?

A. *Ensure your mops are prepared properly, that they are pre-dampened, and that you are using the correct amount of mop heads per sq metre of floor Note; If possible ensure that you machine clean floors as much as possible*

Q. Mops are not effective for sales floor edges

A. *We should still be using doodle bugs to scrub before mopping stubborn edges*

Q. The Easyshine mop tears or rips when I clean around the door metal push plate, why is this happening?

A. *Care should be taken when using Microfibre textiles around any protruding objects such as screws or raised door push plates as this will damage the textiles. Report to your SCM/Supervisor or Handyman if this happens to eliminate any potential injury to customers or colleagues*

Q. What other equipment should I be using on the Sales floor apart from Microfibre?

A. Remember all we have done is replaced a mop for a mop and a cloth for a cloth Note: we should still make use of all other equipment such as doodle bugs, scourers, scrapers, white sponges etc

Bakery/Counters/Restaurant & Prep Areas

Q. The Microfibre Easyshine mop dries before I finish my task*?*

A. Ensure that the mop head is not overused, change frequently as the mop head may dry out quicker within the bakery

Q. My Easyshine mop is too wet what am I doing wrong?

A. Make sure that the water is coming out in an even spray over the mop head rather than a water squirt (adjust the water bottle head to a spray effect)

Q. When cleaning the bakery or counter walls / chilled units or bakery Provers, rack oven surrounding surfaces they sometimes leaves smears

A. *Over the years there will probably have a been build up of chemicals that have been used to clean these areas i.e. degreaser, glass cleaner, general purpose cleaner etc, Microfibre must now work to remove all the build up chemical residue in order for it to leave the areas clean and smear free. Over a short period of time the chemical residue will be removed*

Q. It seems to take longer to remove flour and dough from the bakery mixing bowls using just damp Microfibre cloths

A. We do not need to continue using buckets of water and cloths to remove flour and dough, using the water spray bottle to soften and loosen dough and the yellow Microfibre cloths to remove, works just the

Q. Microfibre does not clean the bakery rack ovens

A. Check the ABC specification for the chemical you can use and the bakery "How to Guide" on how to clean rack ovens, you can also use a "Cleaning Cloth" (a none microfiber cloth) to remove any grease or heavy soiled areas, the Microfibre cloth can be then used as a finishing cloth

Q. The counter / bakery / restaurant floors do not look clean; we are using too many mop heads and taking too much time to clean them

A. if possible all these floor areas should be machine cleaned first, followed by the edges and corners being cleaned with the Microfibre mops i.e. damp 43 or 43 mops and a doodle bug / scourer where required

Q. I cannot reach high level areas to clean e.g. vents, canopies, and conveyor

A. Attached the Easyshine hand tool or spanky to an extendable tool or long handle, to reach high level areas

Q. I have followed the correct process of machine cleaning the floor and damp mopping the corners and edges, but the floor corners and edges around the rack oven and Provers are still dirty, how will I get them clean?

A. You can use a water spray bottle to pre spray floor area, doodle bug, green scourers or cleaning cloth to remove built on grime. If that fails you can use a small amount of de-greaser with any of the above
Shelf Clean

Q. Do I need a Workstation when I clean shelves?
A. If there is one available yes, but in theory you will only need a Utility Belt holding your textiles and equipment or a storage box from the Workstation
Laundering Microfibre Textiles

Q. Textiles are still little damp once they have come out of the drier

A. If the cloths/mops are slightly damp after initial drying process has taken place you can put back into the drier and select the "extra dry button" Note. Textiles should always have an element of residual moisture and not be bone dry.

Q. I washed colour coded cloths together

A. Do not worry, the washer has being programmed to wash all cloths together at a pre-set temperature that kills all germs and bacteria so there is no risk of any cross contamination

Q. **SCM/Supervisor** - if there is a fault with either the washer or drier what should I do?

A. If there is a fault on either the washer / dryer – check the error coding list to verify what the fault is(it may be a slight fault that you can correct i.e. is the detergent bottle empty, is the pipes to washer restricted i.e. blocked pipes. If it isn't something you can rectify report to your Supervisor or SCM who will then log into the Helpdesk

Q. Textiles are coming out of the washer still soiled, what is wrong?

A. It may be a number of things that are causing this problem
- are the textiles being shaken or brushed to remove debris?
- Is the washer over loaded? (The display will show if overloaded)
- Is the detergent pipe at the back nipped or heavy objects have been placed on top of the pipes
- Is there detergent going through?

Q. Do I wash the cloths and mops together?

A. Never, always wash separately

Q. Do the Easyshine mops get washed with the floor mops?

A. No they are washed with the cloths

Q. I cannot remember how many mops and cloths are to be washed and dried at any one time?

A. On both the Laundry bins there should be laminated sheets explaining the amount of cloths and mops to be washed. The maximum cloth load is up to 100 and the maximum mop load is 30

Q. Someone told me that you can put small loads of washing and drying at any one time, is this correct?

A. No, using small amounts of washing / drying wastes energy and is not cost effective, you should wait till a decent size amount of laundry is ready before laundering

Q. Who's responsibility is it to clean the washer / drier and what are needed to be done?

A. It is everybody's responsibility to clean down the washer/drier after use, the following needs to be completed daily:

- Remove fluff from the inner drier door filter
- Remove the drier inner water filter box and drain in the sink
- Damp wipe external surfaces

- Damp wipe rubber seals on both washer & drier doors including drum seals

Q. Are rubber gloves required when loading the washer with textiles?

A. Rubber gloves must be worn when loading textiles into the washing machine – the textiles have been used in toilet areas etc and will be contaminated after use, so wearing gloves is important when handling soiled textiles

Q. **SCM/Supervisor** – I don't know how to fit the Horizon detergent to the washer?

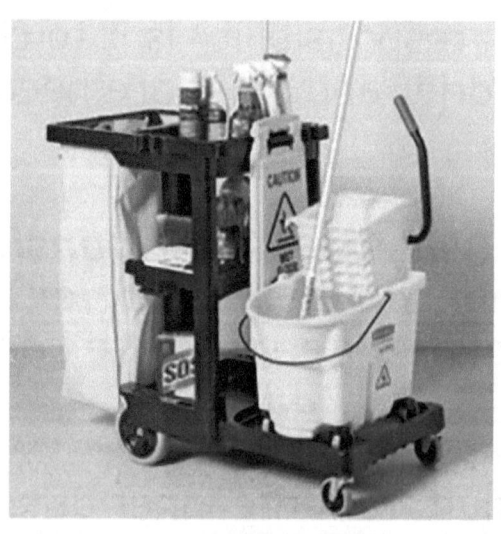

A. From the washer there will be an attached tube, this tube is to be inserted into the bottle cap of the Horizon detergent, there should be a small hole in the cap for the tube to fit into. (if there is no hole in the cap ask your handyman to drill one – there is a small insert on the cap area where he can do this).

When replacing the Horizon detergent bottle swop the bottle cap with the detergent tube onto the new bottle (as any new cap will need the hole drilled out again)

Q. I have been asked by my SCM/Supervisor to wash Hi Vis Vests and I cannot remember how they explained and showed me?

A. Place the vests into the washer making sure you have not exceeded the required load – (no more than 10 at any one time) close the door then turn the dial once, the display unit with show Hi Vis, then press the centre of the dial to operate – refer to the "Washer How To Guide"

CHAPTER FOUR

Key Process

- Introduction of an onsite laundry process within the cleaning room is essential.

  •

- Microfibre systems can be used for areas of cleaning by following due process.

  •

Cleaning Room Set-up as above

- New Washer and Dryer system will be installed and I recommend Electrolux washing system. Simple one programme / one button start is easy to follow
- Significant improvement in standards in cleaning room will be seen – removal of hanging mops (often dirty and smelly) and reduction in malodours

Textiles

- The right cloths & mops have been developed for our specific applications and are reusable to a minimum of 500 washes.

- Once cloth / mop is used and is dirty, the cloth / mop is placed into the dirty compartment of the trolley and replaced with a clean product.

- All cloths and mops are then washed daily and then dried ready for use again.

- Approximately 1.25 days stock of cloths and mops is held within the cleaning room.

Janitor trolley & Light Equipment

- New trolleys & Light Equipment (Mop handles / tools etc) will be issued to your store that have been specifically designed for Microfibre use. These include storage boxes for both clean and dirty mops and cloths.

- All mop tools are securely fixed to the frame of the kit and are generic for all cleaning tasks.

- Mop System is interchangeable for different products / requirements / Colleague height etc.

Impact to Cleaning Standards

- All trial stores have seen improvements in the standards of clean achieved, with areas remaining cleaner for longer.

- Spillages are dealt with easier and more safely as floor is left dry much quicker.

- No more dirty mops and cloths visible on the sales floor.

- Some chemical residue may be visible in areas like toilets until the Microfibre system has removed all traces.

- There is a full detail training programme in this book to support your cleaning teams.

- **Feedback & Questions**

Did You Know?

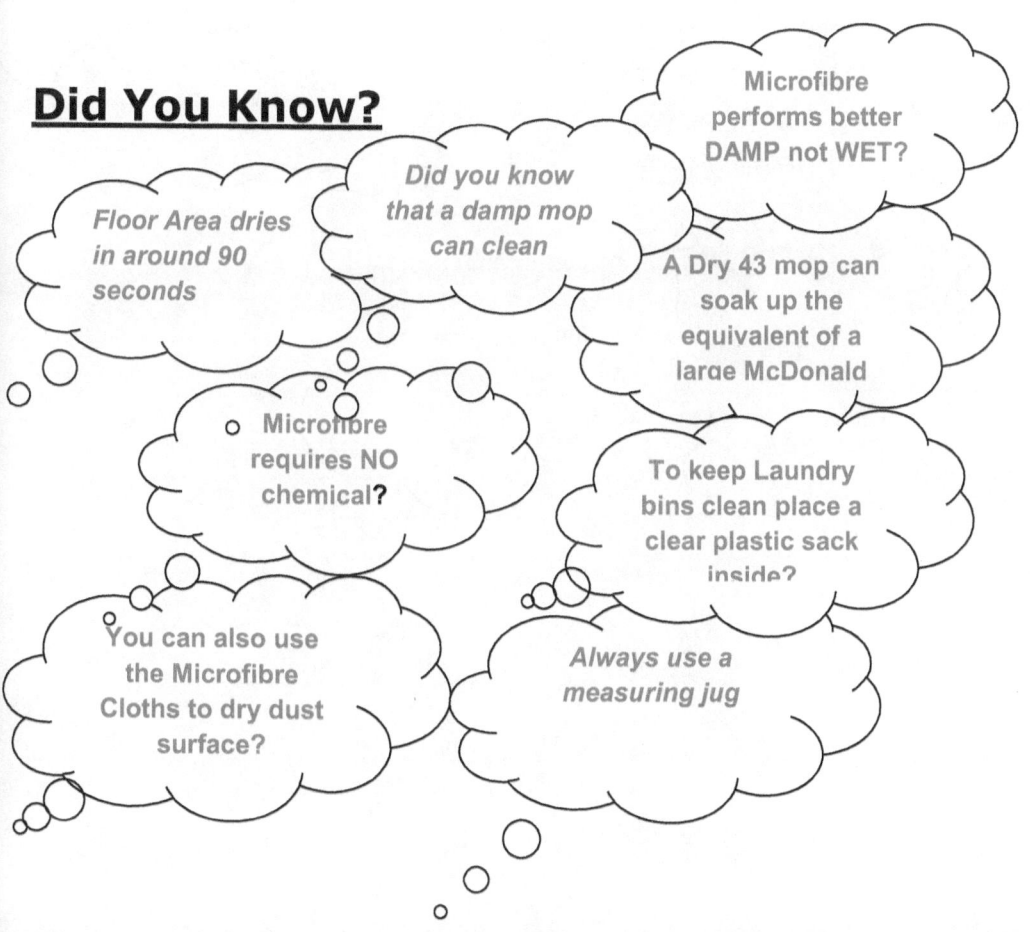

Microfibre performs better DAMP not WET?

Did you know that a damp mop can clean

Floor Area dries in around 90 seconds

A Dry 43 mop can soak up the equivalent of a large McDonald

o Microfibre requires NO chemical?

To keep Laundry bins clean place a clear plastic sack inside?

You can also use the Microfibre Cloths to dry dust surface?

Always use a measuring jug

Did You Know?

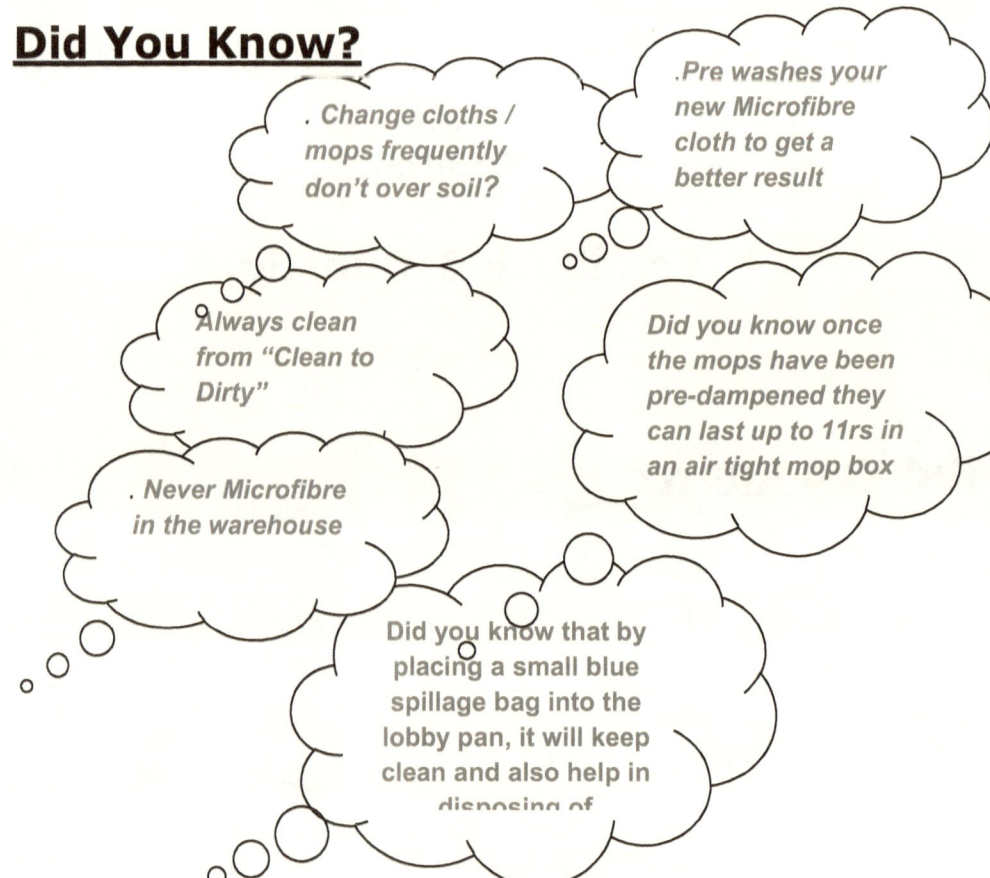

. Change cloths / mops frequently don't over soil?

.Pre washes your new Microfibre cloth to get a better result

Always clean from "Clean to Dirty"

Did you know once the mops have been pre-dampened they can last up to 11rs in an air tight mop box

. Never Microfibre in the warehouse

Did you know that by placing a small blue spillage bag into the lobby pan, it will keep clean and also help in disposing of

ABOUT THE AUTHOR

Joshua Jogo has actively worked in the FM industry since 2002 and has built a continuous incredible weight of experience and expertise in the industry. Coupled with extensive training and capacity building programmes that have made him an authoritative voice in this sector. A highly motivated and enthusiastic individual. Joshua excellent knowledge and experience in cleaning services management and cleaning staff training and supervision. Understanding of cleaning standards and procedures has been a top priority that sets him apart from the rest. He has a track record of inspired achievements which is the result of the best in cleaning quality standards from highly demanding clients. A health and safety champion with Key knowledge of microfiber process and usage. He has written this book to simplify microfiber, process, usage and applications to achieve best results for both business and domestic users. His previous experience with Microfiber at Heathrow airport is an added advantage and you find that knowledge largely implored in this book to help maintain industry standards in microfiber use.

NOTES